THIS BOOK BELONGS TO:

HI! I'M ARTHUR! I love tractors, diggers and trucks (and music). I inspired mama to make this book!

BONUS PLAYLIST
Want to listen to Arthur's favorite tractor, digger, truck and counting songs? Scan the QR code to listen to the book's soundtrack!

COPYRIGHT

Let's Learn Numbers and Counting with Tractors, Diggers and Trucks Copyright ©2025 Ghina Makari. All ideas and content are the author's own. All rights reserved. No part of this book may be reproduced, stored in a retrieval system, or transmitted in any form or by any means (electronic, mechanical, photocopying, recording, or otherwise) without prior written permission of the publisher, except in the case of brief quotations embodied in critical articles or reviews.

Published by ArtyMummaBooks.
@ArtyMummaBooks
ISBN: 978-1-7640580-0-1. First Edition, 2025. Printed by Amazon.

THANK YOU!

Thank you to Dr Michele Dunbar for your expertise and endorsement of this book, the kids committee (Elliot, Rosa and Macy) for their valuable insights; and my husband, family and friends for their support and feedback. To my son Arthur, who inspired this book (and selected every vehicle with love), thank you for being my inspiration and for the joy you bring every day.

CRANE

LET'S LEARN

NUMBERS AND COUNTING WITH

TRACTORS, DIGGERS AND TRUCKS

Ghina Makari Cane

ABOUT SECTION 1

READY TO COUNT?

1, 2, 3... LET'S DIG IN AND LEARN TO COUNT FROM 1-20 WITH TRACTORS, DIGGERS AND TRUCKS!

In this section, your child will:

- boost early math skills by recognizing and counting individual objects
- learn the numbers 1-20 with bright, engaging, learning-focused visuals
- increase vocabulary by learning new words
- get introduced (visually) to the concept of addition (laying the foundation for problem-solving in math, coding and engineering)
- boost STEM learning with fun facts.

1
ONE

DIGGER

2
TWO

CONCRETE MIXERS

3
THREE

STEAM ROLLERS

4
FOUR

CHOMP! CHOMP! I dig deep holes and scoop up dirt! Can you find my long arm?

BACKHOES

5

FIVE

RECYCLING TRUCKS

6
SIX

WEE-OH! WEE-OH! I race to emergencies, carrying water to put out fires! Can you find my ladder?

FIRE TRUCKS

7

SEVEN

TRACTORS

8
EIGHT

BULLDOZERS

9
NINE

FORKLIFTS

10
TEN

CRASH! CLUNK! I carry loads of dirt and dump them with a big thud! What color am I?

DUMP TRUCKS

11 ELEVEN

12 TWELVE

13 THIRTEEN

14 FOURTEEN

15 FIFTEEN

16 SIXTEEN

17 SEVENTEEN

18 EIGHTEEN

19 NINETEEN

20 TWENTY

ABOUT SECTION 2

READY FOR A CHALLENGE?

BEEP BEEP! LET'S DO SOME FUN CHALLENGES WITH THE HELP OF PLANES, CRANES, CARS AND MORE VEHICLES.

In this section, your child will:

- apply learnings (count by themself)
- boost critical thinking and problem solving (essential for math, engineering and coding)
- learn to distinguish between different objects
- get introduced to the mathematical concepts of more than, fewer than and equal to, as well as, odd and even numbers
- get introduced to the concepts of size, speed, weight and mass.

COUNT THE NUMBER OF OBJECTS IN EACH SECTION.

HELICOPTERS

AIRPLANES

COUNT THE NUMBER OF OBJECTS IN EACH SECTION.

COUNT THE NUMBER OF MONSTER TRUCKS.

SCHOOL BUSES

MONSTER TRUCKS

COUNT THE NUMBER OF CRANES.

CRANES

POLICE CAR

ARE THERE MORE DIGGERS OR BACKHOES?

BACKHOE

DIGGERS

ARE THERE MORE AUGER EXCAVATORS OR TRUCKS?

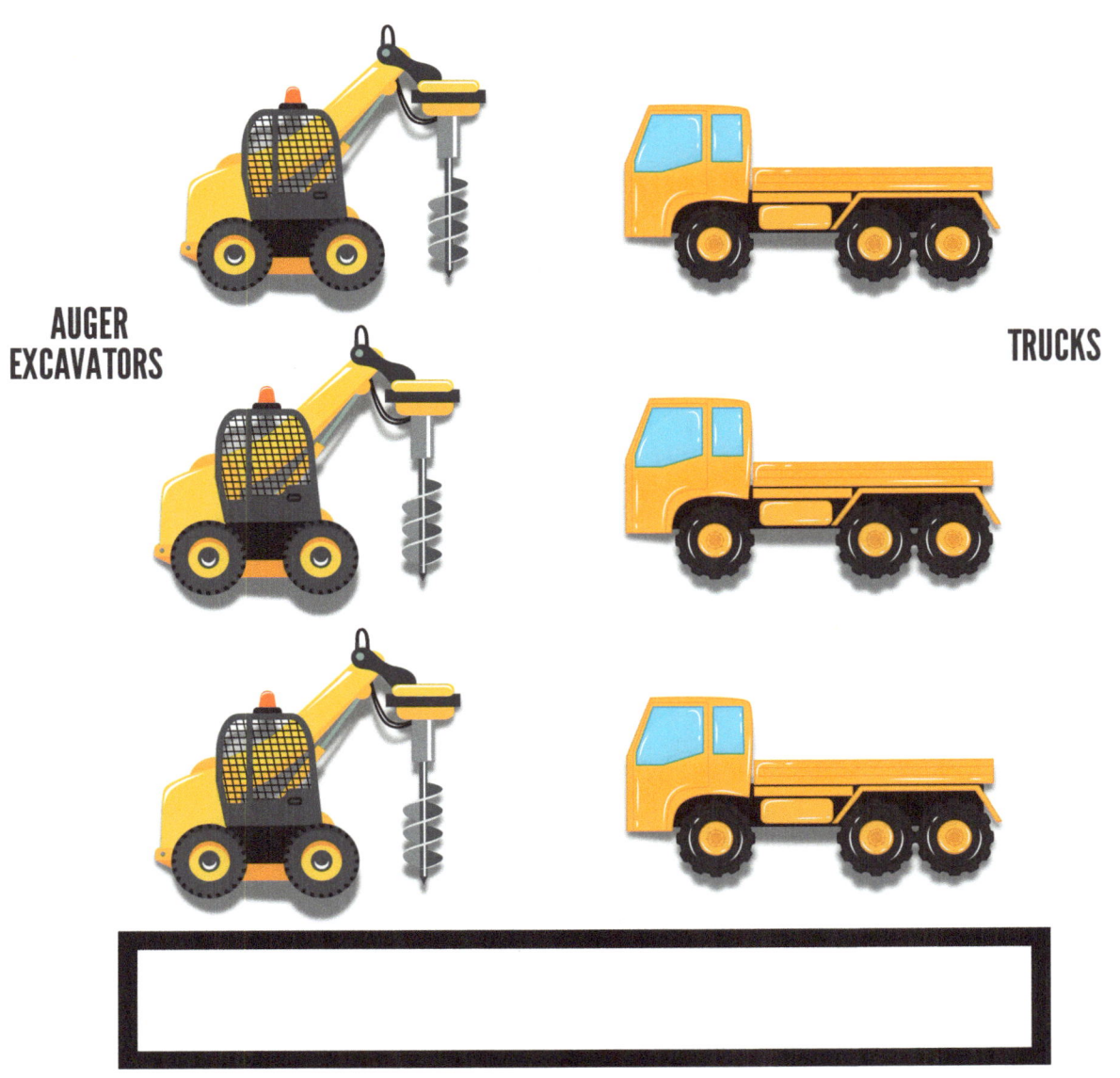

WHICH OBJECT IS BIGGER?

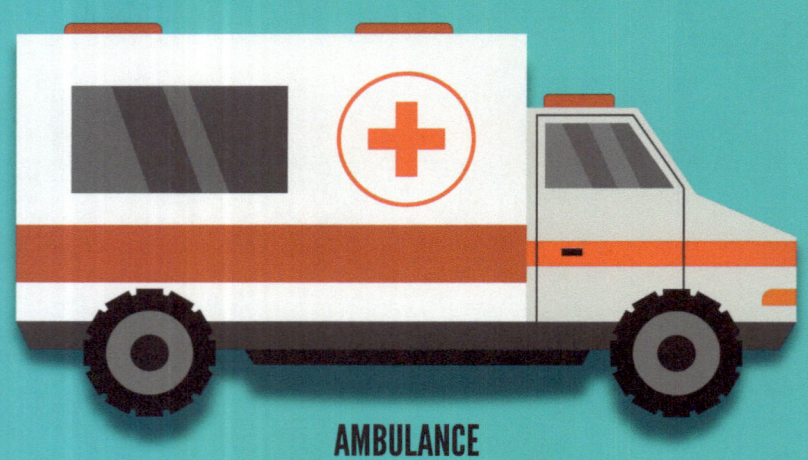

AMBULANCE

MOTOR BIKE

WHICH OBJECT CAN GO FASTER?

PLOUGH TRACTOR

RACE CAR

WHICH OBJECT IS HEAVIER?

CONCRETE MIXER

BICYCLE

WHICH DUMP TRUCK HAS MORE DIRT? THE TOP OR BOTTOM ONE?

DUMP TRUCKS

ABOUT SECTION 3

DIGGING DEEPER!

LET'S KEEP LEARNING: A DISCUSSION GUIDE FOR PARENTS, CAREGIVERS & EDUCATORS.

Reading's only half the fun—talking, wondering, and thinking makes it even better!

Here are some fun questions you can ask every time you read together (you don't have to ask them all, just pick a few each time!). There are no right or wrong answers (just big imaginations at work).

Each time you read, your child will notice something new helping to build early math, coding, engineering and problem solving skills... one truck (or digger!) at a time.

QUESTIONS

1. **How many wheels can you spot?** *Big trucks sometimes have a lot! Count on one page or count them all!*

2. **Which machine do you think sounds the loudest?** *No wrong answers, just big imaginations at work!*

3. **Did you notice when there was just one object... or lots of objects on the page?** *Did you spot the 's' sound when there was more than one?*

4. **What colors did you spot today?** *How many yellow vehicles did you see? How many green trucks?*

5. **What's something new you learned from the book today?** *A new machine? A new number? A fun fact? A new word?*

6. **If you could drive one of these vehicles, which one would you pick? Where would you go?** *Dream big and let's go on an adventure!*

7. **Which page did you enjoy the most and why?** *Maybe it's the noisiest truck... or the biggest wheels!*

ABOUT THE AUTHOR

A PROUD MAMA AND PASSIONATE CREATOR.
AUTHOR | ARTIST | MAMA

Ghina Makari Cane (@ghinamak) is a proud mama, writer and artist, and an award-winning marketer with over 20 years' experience. She has turned her creative flair to children's books to fulfill a lifelong dream of becoming an author.

This is her debut book, inspired by her toddler, who said, "Mama, Mama, I want four diggers and five trucks!" while she was working on a zoo book. And just like that, the idea for a fun and educational counting book was born! Look out for more books in the 'Let's Learn' series including a coloring book and multilingual editions.

Ghina is the founder of ArtyMummaBooks, an independent publisher on a mission to make learning and languages fun, one book at a time.

 | @artymummabooks

FREE POSTERS!

Want a free numbers and counting poster pack?

Poster pack includes a full-color and coloring-in poster!

SCAN THE QR CODE TO GET YOUR FREE POSTER PACK.

Thank you readers!

Thank you for reading this book. It would be amazing if you could take a moment to **review this book on Amazon**. Thanks in advance!

To purchase copies of this book, please visit Amazon.

WRECKING BALL CRANE

www.ingramcontent.com/pod-product-compliance
Lightning Source LLC
Chambersburg PA
CBHW041103070526
44583CB00002B/41